本书由中国科学院数学与系统科学研究院资助出版

数学 24/7

计算机中的数学

〔美〕詹姆斯·菲舍尔 著

龙 静 译

科学出版社

北 京

图字：01-2015-5626号

内 容 简 介

计算机中的数学是"数学生活"系列之一，内容涉及下载速度、上传速度、计算机存储、二进制、十六进制、网页设计等方面，同时介绍了不同计量单位之间的换算、二进制与十六进制之间的换算等知识，让青少年在学校学到的数学知识应用到与计算机有关的多个方面中，让青少年进一步了解数学在日常生活中是如何运用的。

本书适合作为中小学生的课外辅导书，也可作为中小学生的兴趣读物。

图书在版编目（CIP）数据

计算机中的数学/（美）詹姆斯·菲舍尔（James Fischer）著; 龙静译.—北京:
科学出版社,2018.5
（数学生活）
书名原文：Computer Math
ISBN 978-7-03-055693-6

Ⅰ.①计⋯ Ⅱ.①詹⋯ ②龙⋯ Ⅲ.①数学-青少年读物 Ⅳ.①O1-49

中国版本图书馆CIP数据核字（2017）第292430号

责任编辑:胡庆家 / 责任校对:邹慧卿
责任印制:肖 兴 / 封面设计:陈 敬

科 学 出 版 社 出版
北京东黄城根北街16号
邮政编码：100717
http://www.sciencep.com

北京汇瑞嘉合文化发展有限公司 印刷
科学出版社发行 各地新华书店经销
*
2018年5月第 一 版 开本:889×1194 1/16
2018年5月第一次印刷 印张:4 1/4
字数:70 000
定价：98.00元(含2册)
（如有印装质量问题，我社负责调换）

引 言

你会如何定义数学？它也许不是你想象的那样简单。我们都知道数学和数字有关。我们常常认为它是科学，尤其是自然科学、工程和医药学的一部分，甚至是基础部分。谈及数学，大多数人会想到方程和黑板、公式和课本。

但其实数学远不止这些。例如，在公元前5世纪，古希腊雕刻家波留克列特斯曾经用数学雕刻出了"完美"的人体像。又例如，还记得列昂纳多·达·芬奇吗？他曾使用有着赏心悦目的尺寸的几何矩形——他称之为"黄金矩形"，创作出了著名的画作——蒙娜丽莎。

数学和艺术？是的！数学对包括医药和美术在内的诸多学科都至关重要。计数、计算、测量、对图形和物理运动的研究，这些都被融入到音乐与游戏、科学与建筑之中。事实上，作为一种描述我们周围世界的方式，数学形成于日常生活的需要。数学给我们提供了一种去理解真实世界的方法——继而用切实可行的途径来控制世界。

例如，当两个人合作建造一样东西时，他们肯定需要一种语言来讨论将要使用的材料和要建造的对象。但如果他们建造的过程中没有用到一个标尺，也不用任何方式告诉对方尺寸，甚至他们不能互相交流，那他们建造出来的东西会是什么样的呢？

事实上，即便没有察觉到，但我们确实每天都在使用数学。当我们购物、运动、查看时间、外出旅行、出差办事，甚至烹饪时都用到了数学。无论有没有意识到，我们在数不清的日常活动中用着数学。数学几乎每时每刻都在发生。

很多人都觉得自己讨厌数学。在我们的想象中，数学就是枯燥乏味的老教授做着无穷无尽的计算。我们会认为数学和实际生活没有关系；离开了数学课堂，在真实世界里我们再不用考虑与数学有关的事情了。

然而事实却是数学使我们生活各方面变得更好。不懂得基本的数学应用的人会遇到很多问题。例如，美联储发现，那些破产的人的负债是他们所得收入的1.5倍左右——换句话说，假设他们年收入是24000美元，那么平均负债是36000美元。懂得基本的减法，会使他们提前意识到风险从而避免破产。

作为一个成年人，无论你的职业是什么，都会或多或少地依赖于你的数学计算能力。没有数学技巧，你就无法成为科学家、护士、工程师或者计算机专家，就无法得到商学院学位，就无法成为一名服务生、一位建造师或收银员。

体育运动也需要数学。从得分到战术，都需要你理解数学——所以无论你是想在电视上看一场足球比赛，还是想在赛场上成为一流的运动员，数学技巧都会给你带来更好的体验。

还有计算机的使用。从农庄到工厂、从餐馆到理发店，如今所有的商家都至少拥有一台电脑。千兆字节、数据、电子表格、程序设计，这些都要求你对数学有一定的理解能力。当然，电脑会提供很多自动运算的数学函数，但你还得知道如何使用这些函数，你得理解电脑运行结果的含义。

这类数学是一种技能，但我们总是在需要做快速计算时才会意识到自己需要这种技能。于是，有时我们会抓耳挠腮，不知道如何将学校里学的数学应用在实际生活中。这套丛书将助你一马当先，让你提前练习数学在各种生活情境里的运用。这套丛书将会带你入门——但如果想掌握更多，你必须专心上数学课，认真完成作业，除此之外再无捷径。

但是，付出的这些努力会在之后的生活里——几乎每时每刻（24/7）——让你受益匪浅！

目　　　录

Contents

1
下载速度

特雷莎一家刚搬到一座新城市。她正在整理自己的房间，父母在整理其他房间。

他们忙于布置房间，还没有开始使用电脑。当特雷莎的哥哥阿杰打开笔记本电脑，让爸爸设置家庭网络时，才发现无法接入互联网。特雷莎的父母为搬家做了充分的准备，却忘了为新家申请互联网服务。

特雷莎和爸爸去了当地的图书馆，那里有很多电脑并可以免费上网。他们查找最好的互联网服务，以便打电话订购。

特雷莎看到很多数字，不明白什么意思，例如 512/128 Kb/s。爸爸说那分别表示下载和上传速度。第一个数字是下载速度，下载是将信息从其他计算机传送到你的计算机上，比如可以下载音乐、电子表格、电子邮件的附件，等等。

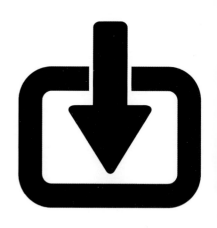

描述下载速度没有一个标准的表示，人们通常使用Kb/s 或Kbit/s表示千比特每秒，Mb/s 或 Mbit/s则表示兆比特每秒。

当你试图下载一个67Mb的文件时，以更高的速度下载将更快，如相比6Mb/s，而以512Kb/s的速度下载则比较慢。

计算一个文件的下载时间需要以下步骤：

• 通过除以8将下载速度从千比特/秒或兆比特/秒转换成千字节/秒或兆字节/秒（下文用b，Kb，Mb分别表示比特、千比特、兆比特，用B，KB，MB，GB分别表示字节、千字节、兆字节、千兆字节）。（此步骤将在第3节有更多解释。）

• 确保文件大小的单位和下载速度的单位一致。否则，请按如下换算方式转换：

$$1024 \text{ B} = 1\text{KB}$$
$$1024 \text{ KB} = 1 \text{ MB}$$
$$1024 \text{ MB} = 1 \text{ GB}$$

• 将文件大小除以下载速度得到下载需要的秒数。

计算以 6 Mb/s的速度下载980 KB的文件所需时间：

$$6 \text{ Mb/s} \div 8 = 0.75 \text{ MB/s}$$
$$980 \text{ KB} \div 1024 = 0.96 \text{ MB}$$
$$0.96 \text{ MB} \div 0.75 \text{ MB/s} = 1.28 \text{ s}$$

1. 以 512 Kb/s 的速度下载这个文件需要多少时间？

2. 两种下载速度所需时间是否都超过1秒?你会注意到这两种速度的差别吗?

2
上传速度

现在特雷莎明白了下载速度，她还想了解上传速度。还记得互联网服务包提供的第二个数字吗？512/128 Kb/s 中第二个数字是上传速度。上传是指从你的计算机传输到其他计算机。你可以将图片上传到 FACEBOOK 或将视频上传到 YOUTUBE。

特雷莎和爸爸需要弄清楚他们想要多快的上传速度。一旦确定，就可以选择合适的互联网服务包。

上传速度往往要比下载速度慢。但是，如同下载速度一样，你可以计算同一文件在不同上传速度时所需的上传时间。

1. 如果要上传 12 MB 的文件，通过以下上传速度分别需要多少秒：

560 Kb/s:

2 Mb/s:

1 Mb/s:

128 Kb/s:

5 Mb/s:

现在特雷莎明白，更快的上传和下载速度意味着可以在计算机中更快地添加音乐，更流畅地看电视，她和爸爸都认为不要最慢的。但是，他们到底需要多快的网速呢？

有如下互联网服务包供选择：

标准包：15/1 Mb/s，40 美元/月
特级包：30/5 Mb/s，50 美元/月
超级包：50/5 Mb/s，65 美元/月

因为特雷莎的妈妈有时需要做一些平面设计的工作，需要下载或上传的文件最大约 20 MB，这相当大了。

2. 上传、下载 20MB 文件，下面每个互联网服务包分别有多快？

标准包：
特级包：
超级包：

3. 特雷莎的父母不想每月花费超过 50 美元。他们应该选择哪个互联网服务包？如果他们不订购超级包，会注意到速度减慢吗？

3
计算机存储

特雷莎和爸爸回家了，但是今晚打电话给互联网公司太晚了——下班了。特雷莎坐在计算机旁，但不知道做什么。以前她总是浏览网页或在线观看喜欢的电视节目。

特雷莎随意点击着计算机上以前没有用过的按钮。突然，她看到一个界面显示计算机有多少存储空间，其中有兆字节、千兆字节这样的字眼，与下载和上传速度中的千比特、兆比特看起来非常类似。她叫来哥哥阿杰，哥哥解释说字节、千字节、兆字节都是用来描述一台计算机有多少存储空间的。不像上网速度以比特为单位，计算机存储以字节为单位。有意思的是，这两者是相关的。

比特是计算机最小的信息量单位。字节稍大一点——1字节等于8比特。

如何描述计算机上更大的信息？千字节比字节大，兆字节比千字节大，以此类推。下面的图表显示了这些计量单位之间的关系：

$$8b = 1 B$$
$$1024 B = 1 KB$$
$$1024 KB = 1 MB$$
$$1024 MB = 1 GB$$
$$1024 GB = 1 TB$$
$$1024 TB = 1 PB$$
$$1024 PB = 1 EB$$

1. 1 KB等于多少GB？

2. 3 EB等于多少TB？

计算得到的数字相当大！
当人们在谈论互联网速度时说到Kb或Mb，实际上说的是KB或MB除以8。

3. 89字节等于多少比特？

4
电子邮件存储

特雷莎家里安装了互联网后，她做的第一件事就是检查电子邮件。她想看看是否收到了来自老朋友的邮件。

看完所有新邮件后，她注意到屏幕的角落上有一个小长条，显示已使用80%。她之前从来没有真正注意到，因此仔细看了看。

果然，她看到"10.1 GB 已使用 80%"。居然占用了这么多空间！她早就知道计算机有存储空间，但现在她才知道邮件和附件也占用存储空间，下面我们计算一下特雷莎已经使用多少存储空间，还剩多少。

特雷莎的邮箱共有10.1GB的空间。要计算还剩多少存储空间得使用百分数。

要计算10.1GB的80%，先向左移动小数点两位，将百分数转换为十进制数，80%=0.80。将这个数乘以10.1GB就可以知道特雷莎使用了多少空间。

1. 特雷莎使用了多少存储空间？还剩多少？

这个问题还有另一种解法，即将百分数转换成分数：

$$80\% = \frac{80}{100}$$

这个分数等于已使用的空间除以总空间。假设已使用的存储空间是 x，则

$$\frac{80}{100} = \frac{x}{10.1}$$

2. 交叉相乘后得到多少？是否和之前结果一样？

3. 她使用了多少KB？

为了给新邮件腾出更多存储空间，特雷莎决定删去一些旧邮件。仔细检查后，得知删了1924MB旧邮件。

4. 她的邮箱现在用了多少GB？占用空间的百分比是多少？

5

二进制代码

特雷莎想知道比特、字节、千兆字节究竟是什么，她知道这是计算机中信息的度量单位，但不确定它们是如何度量的？1 比特到底有多大？她向阿杰问了这些问题。

阿杰告诉她需要了解二进制代码，这是计算机使用的语言。计算机无法理解人类使用的字母和数字，而是将信息存储为包含 0 和 1 的字符串。二进制"只有两种状态"，计算机使用的二进制代码只有 1 或 0。

计算机使用这些 0 和 1 的不同组合来表示其他字符。更多讨论请见第 6 节。

比特实际上是二进制位的英文缩写。一个二进制位是计算机存储信息的最小单位。在计算机语言中一个二进制位是 0 或 1。你可以说 1 等于"开"或"是"，0 等于"关"或"非"。

二进制的计数系统不同于通常的计数系统。我们通常使用 由 10 个数字 (0 到 9) 组成的十进制计数系统。在二进制代码中只有两个数字——但需要计数的值比 2 大许多！

首先，需要了解2的幂。幂是一个数乘以自身的次数。2 的 0 次幂 $2^0=1$， 2 的 1 次幂 $2^1=2$，

$$2^2 = 2 \times 2 = 4$$
$$2^3 = 2 \times 2 \times 2 = 8$$
$$2^4 = 2 \times 2 \times 2 \times 2 = 16$$

1. 2^7 等于多少？

数字 0 到 10 写成二进制是这样的：

0000 (= 0)   0110 (= 6)
0001 (= 1)   0111 (= 7)
0010 (= 2)   1000 (= 8)
0011 (= 3)   1001 (= 9)
0100 (= 4)   1010 (= 10)
0101 (= 5)

在二进制系统中计数，总是从 0 开始，接下来是 1。但之后呢？回想一下 2 的幂。每个数位都分派一个 2 的幂，数位上是 1 表示"开"，0 表示"关"。然后将所有含 1 的列加起来。比如 10（在十进制中等于 2），将 10 分成两列。右边的列是 2^0，左边的列是 2^1。只有左边是"开"，因为它是 1。这是唯一要计数的列。2^1 是 2，这就是答案。

对于 3，两列都是"开"，从右到左，可以得到 $2^0 + 2^1 = 3$。对于 4，只有 2^2 这列，因为只有这列是"开"，$2^2=4$。

2. 如何解释数字 8 和 9？

一个字节有 8 位，取值范围从 00000000 到 11111111（对于 1 或 0，有 8 个选择）。

3. 字节 11001000 的取值是多少？

6
如何编写二进制代码

阿杰继续解释二进制、比特和字节，以及它们的真正含义。他让特雷莎打开写字板（计算机上最基本的文本编辑程序），并打出她的名字。"你看到了什么？"阿杰问她。特雷莎说她只看到组成她名字的字母。但是阿杰说，对于计算机而言，这些字母就像二进制代码。计算机只能识别包含1和0的字符串，不能识别字母。计算机将屏幕上的每个字符识别为1个字节（8位）。然后阿杰告诉她如何编写二进制代码，而不用任何字母。下面教你如何做。

屏幕上的每个字符 (字母或符号，比如一个句号或破折号) 是计算机上 1 个字节的信息。

人们提出一个系统，里面每个字符被分配一个特定的字节代码。例如

$$A = 01000001$$
$$B = 01000010$$
$$C = 01000011$$

1. 如果将 A, B, C 转换为正常的（十进制）数，它们是什么？

当特雷莎写出她的名字，用二进制表示的字节信息是

$$T = 01010100$$
$$E = 01100101$$
$$R = 01110010$$
$$E = 01100101$$
$$S = 01110011$$
$$A = 01100001$$

2. 每个字母的十进制数是多少？

3. 如果特雷莎的姓以字节为72的字母开头，那么是下面哪个字母？

$$M = 01001101$$
$$D = 01000100$$
$$H = 01001000$$

7
十六进制的转换

阿杰告诉特雷莎，除了二进制，计算机还能理解另一种计数系统：十六进制。就像二进制数以 2 为基数，通常使用的十进制数以 10 为基数，十六进制数则以 16 为基数。我们的输入系统仅有十个数字，这显然不够用！如果计算机从 9 以后开始使用字母来表示，这就足够了。下面来看十六进制中如何表示数字1到16：

1 = 1	9 = 9
2 = 2	A = 10
3 = 3	B = 11
4 = 4	C = 12
5 = 5	D = 13
6 = 6	E = 14
7 = 7	F = 15
8 = 8	10 = 16

二进制和十六进制间的转换很容易。由于$16=2^4$，每四位二进制数对应于一位十六进制数。

二进制	十进制	十六进制
0001	1	1
0101	5	5
1010	10	A
1011	11	B
1100	12	C
1101	13	D
1111	15	F

1. 二进制数 1001 的十六进制表示是多少？

2. 十六进制数 1A 的十进制表示是多少？

3. 十六进制数 C1 的二进制表示是多少？

Google

Google Search | I'm Feeling Lucky

8
计算机逻辑

进 入新学校后，特雷莎开始做一项研究。她要研究小镇的历史。她真的不了解这个新的小镇，所以需要做大量的调查。

特雷莎开始使用谷歌进行搜索。首先，她搜索"布鲁克维尔"——小镇的名字。她得到了很多结果，但是大部分与小镇无关。她希望只是搜索到想要的结果。如果可以精炼搜索结果，研究会容易得多。

幸运地是，有一个简单的方法可以做到这一点。她可以使用布尔搜索。布尔逻辑包含条件"与""或"和"非"。布尔搜索是计算机使用的逻辑系统的一部分。谷歌搜索时会使用它，计算机也使用它处理日常事务。

开始，特雷莎仅仅只会在搜索中添加一项。布鲁克维尔在纽约州，因此她可以搜索"布鲁克维尔纽约"。她没有使用"与"，因为谷歌默认使用"与"。这样搜索结果包含 布鲁克维尔 和纽约两个词。

特雷莎注意到她的搜索结果包含很多其他州的布鲁克维尔，特别是得克萨斯州。她可以搜索"布鲁克维尔-得克萨斯州。"在谷歌搜索条件中，得克萨斯前面的减号意味着"非"。现在，她的搜索结果中排除了得克萨斯州一词。

这个小镇很久以前叫"布鲁克顿"。特雷莎通过搜索"布鲁克维尔"或"布鲁克顿"可以同时找到布鲁克维尔和布鲁克顿。"或"表示包含一方或其他。

计算机也使用这种逻辑，只是不是以词汇形式。计算机使用字符串 1 和 0，可以使用表格来显示这种逻辑关系。下面的表给出了"与"命令及其说明。在这个表中，A 是第一个输入，B 是第二个输入，而 Q 是输出。只有 A 和 B 都为 1 时，Q 才为1。

```
A B Q
0 0 0    若 A=0 与 B=0，则 Q=0.
0 1 0    若 A=0 与 B=1，则 Q=0.
1 0 0    若 A=1 与 B=0，则 Q=0.
1 1 1
```

1. 在"与"命令表中"111"表示什么？

"或"命令表类似。如果 A=1 或 B=1，那么输出为1。请填写下表中其他的说明。

2.
```
A B Q
0 0 0    若 A=0 与 B=0，则 Q=0.
0 1 1
1 0 1
1 1 1
```

下面是"非"命令表。只有一个输入，而输出总是相反的。

```
A Q
0 1
1 0
```

3. 特雷莎搜索的"布鲁克顿-得克萨斯州"在计算机术语言中看起来像什么？可以看作"布鲁克顿与非得克萨斯州"，它排除了"布鲁克顿与得克萨斯州"的搜索结果。

```
        get { return _host;
}
public DialogResult ShowDialog(IWin32Window owner, Host
{
        _host = host;
        DialogResult res = ShowDialog(owner).
        if (res == DialogResult.OK)
        {
            if (_host == null)
            {
                    bool ne =
sNullOrEmpty(_tbHostName.Text);
                    bool ae =
sNullOrEmpty(_tbHostIp.Text);
                    ne && ae)
```

9
简单的编程

特 雷莎在新学校认识了一些好朋友。她的一个新朋友梅尔碰巧擅长使用计算机。

一天，梅尔在特雷莎家做客，特雷莎提到她的计算器电池用完了，还没来得及换新的。梅尔告诉她不要担心，可以把计算机变成一个计算器。

梅尔解释说，特雷莎需要先学习一点编程。编程就是告诉计算机指令，有点像与计算机沟通的另一种语言。计算机将程序命令（又叫程序代码）翻译成二进制数，以此来识别你要它做什么。梅尔已经学会了一点编程知识，所以她能告诉特雷莎怎么做。

计算机有数百种不同的编程语言。只要遵守它的语言规则，不管用哪种都可以。本节的示例都是已知的程序语言。你知道或了解的计算机语言会稍有不同，但它们在做同样的事情。

梅尔告诉特雷莎所有的计算机程序都由一行一行的代码构成。每一行是一个命令。程序要告诉计算机什么时候开始运行，什么时候结束，比如：

```
1 Program
2 End Program
```

在这两个命令之间添加指令就能将计算机变成计算器，或运行其他程序。

特雷莎的数学作业涉及大量的立方数（数字乘以自身3次）。回想一下二进制数和2的幂。2的立方是2^3或$2 \times 2 \times 2$。特雷莎希望她的计算机程序能够计算数字的立方。

按照如下步骤，记住，开始是"Program"，最后是"End Program"：

首先，在写入立方的代码之前，必须告诉计算机设置空间来存储数字和它们的立方数。这就是所谓的变量声明。我们将变量命名为"xCubed"。可以通过存储信息的类型（文字或数字）来创建一个新的变量并命名它。例如，"Number(xCubed)"将创建一个名为"xCubed"的数字变量。

接下来，梅尔告诉特雷莎她必须告诉计算机求 2 的立方是2乘以自身三次，并把所得到的值放入"xCubed"。

最后，特雷莎告诉计算机在第7行给出求立方的答案。可以将答案用字符打印输出。

重新排列计算机中的下面这些代码行，告诉计算机如何计算平方根。括号外面的字符是命令，括号内的是可修改的参数。

```
Print "This program will find the result of the
equation 2³."
End Program
Program
Number (xCubed)
Print "Your number cubed is ", (xCubed)
(xCubed= 2 * 2 * 2 )
```

10
网页设计：像素

特雷莎想要创建个人主页，用来与她原来城镇的朋友进行交流。她总是太忙，没有时间给每个人打电话或发邮件。更新个人主页是一个与所有朋友保持联系的很好的方法。

特雷莎使用了比较容易上手的网页设计程序。首先需要决定将网页设计成什么样。例如，需要选择适当的尺寸，页面不能太大或太小。

网页以像素为度量单位，就像其他东西以英寸、厘米或英里来度量。一个像素是计算机屏幕图像的一个小点。一幅图像由成千上万的像素组成。像素是如此之小，以至于你看不到它们——但它们融合在一起就组成了一个平滑的图片。一个美观的网页应该有合适的尺寸。

当选择网页的像素时，你可能面临如下选项：

640 × 480
1024 × 768
800 × 600

这里640×480是指屏幕宽度640像素，高度480像素。这个数字叫分辨率。

计算网页的分辨率就像计算面积(将长度与宽度相乘)一样。

1. 640×480的网页分辨率是多少？

2. 1024×768和1280×800的网页，哪个分辨率更大？大多少？

一个好的网页分辨率能适合大多数人的计算机屏幕。如果你的计算机分辨率是800×600，当浏览1024×768的网页时，网页就显得太大了。你将无法用整个屏幕看到它的全部，而不得不滚动到右侧以查看整个网页。

特雷莎对她的朋友们做一项民意调查，了解他们的计算机屏幕的分辨率是多少，这样她就可以为她的网页挑选合适的分辨率。调查结果是：

800 × 600 = 2人
1024 × 768 = 6人
1280 × 800 = 1人

3. 特雷莎应该选择哪个分辨率？为什么？

11
网页设计：色彩

下一步，特雷莎要为她的个人主页挑选一些颜色。当她在网页设计程序中挑选颜色时，注意到每种颜色有一个有趣的编号作为标识，包含一堆数字，也有一些字母。比如，紫色是C030FF。它是什么意思呢？

特雷莎考虑用十六进制方式进行颜色设置。类似于二进制，十六进制只是一种不同的计数方式。我们通常使用的是基于10的十进制。二进制基于2，十六进制基于16。十六进制有十个数字0到9，剩下的使用A到F这六个字母。

要获得一种颜色的十六进制代码，需要组合3个不同的十六进制数字。紫色中的C0表示有多少红色；30表示有多少绿色；FF表示有多少蓝色。

首先需要更好地理解如何在十六进制中计数。

十六进制中前16个数是0，1，2，3，4，5，6，7，8，9，A，B，C，D，E，F。随后是

$$10 (= 16)$$
$$11 (= 17)$$
$$12 (= 18)$$

这是怎么回事？回想二进制数，它的每列都对应 2 的幂。十六进制数也是一样的，只是对应的是16的幂。

十六进制的10意味着 $1 \times 16^1 + 0 \times 16^0$，等于16.

1. 在十六进制中，14等于多少？

你需要记住A到F分别等于多少。

4C实际上是 $4 \times 16^1 + 12 \times 16^0$。在十六进制中C是12，因此 $4C = 64 + 12 = 76$。

对于颜色，计算机按顺序将红色、绿色和蓝色混合在一起。绿色已经包含黄色，因此计算机可以得到黄色。每种颜色都有两位数的十六进制代码。这就是为什么每种颜色其实有6位数，因为它结合了3种颜色。第一组两个数表示有多少红色，第二组两个数表示有多少绿/黄色，第三组两个数字表示有多少蓝色。

2. E41D14是什么颜色？

数值越低，颜色越暗(更接近黑色)；数值越高，越接近白色：000000是黑色，而 FFFFFF 是白色。

3. 蓝色值E9会更接近于黑色还是白色？ 为什么？

特雷莎选择的紫色是C030FF。现在考虑16的幂，0到5，因为有6列数字。

4. 在十进制中，紫色对应的数值是多少？

12
网页设计：布局

特雷莎的个人主页做好了一半，但她还需要添加图像和文本框。她将所有需要包含的东西列了一张表，但要确保它们适合于每个网页。为此，她反复调试每个图像和文本框的大小。

特雷莎很快就发现她在与比例打交道。例如，当她更改图像大小时，可以选择按比例调整。当她将图像调宽时，就同时将它变长，这样使图像看起来比较正常，只是会更大。她也可以不按比例调整，只是会变得太宽或太长，这使它看起来被压扁了。了解比例会使网页设计更容易，也更好看。

成比例的两个事物具有相同的比例关系。3英寸×6英寸的图片的宽度(3英寸)和高度(6英寸)之间有具体的比例关系。可以用分式表示这种关系：

$$\frac{宽度}{高度}$$

$$\frac{3英寸}{6英寸}$$

特雷莎希望网页上的图片大于3英寸×6英寸。然而，她想让它看起来与原图一样并且有相同的比例，而不会被压扁。

事实上，她希望这张图片是5英寸宽。要找到新的高度，必须找到一个分式，它与原始图片宽度和高度之间有相同的比例关系，这个关系是3/6。

使两个图片大小比例相等，不考虑图像清晰度。假设新的高度为X，

$$\frac{3}{6} = \frac{5}{X}$$

现在交叉相乘：

1.

$$3 \times X = 5 \times 6$$
$$X =$$

2. 如果特雷莎想要图片的高度小一英寸，宽度和高度新的比例是什么？

3. 下面哪个屏幕分辨率与1024×768成比例？（答案可能不止一个）

    A. 960 × 720
    B. 976 × 580
    C. 1080 × 925
    D. 1140 × 855

13

Excel

特雷莎的计算机上有EXCEL软件，在学校课程中她经常使用这个软件。她甚至在之前做历史课题（关于她居住的新城镇的历史）时也使用EXCEL软件。

特雷莎的课题的第二部分是调查这些年来共有多少人居住在小镇上。通过搜索，特雷莎记下她能找到的每年的数字，但她希望能做进一步的工作。她将所有的信息输入EXCEL中，并制作成一个图表反映小镇的人口。请看下一页的图表。

这是特雷莎输入到EXCEL中的图表：

年份	人口
1750	45
1775	231
1800	879
1825	1670
1850	4006
1875	7096
1900	10213
1925	10607
1950	10890
1975	9992
2000	8863
2025	?

1. 特雷莎的图上的 x 轴代表什么？y 轴呢？

请在下图的数轴上添加标记。

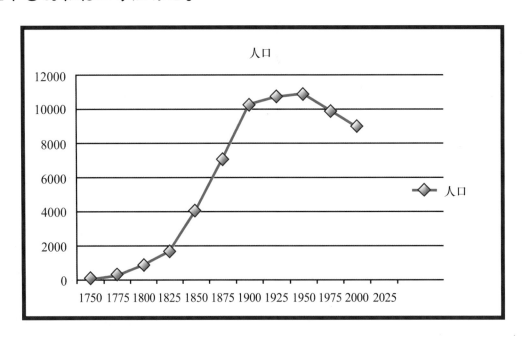

通过这个图，可以估计2025年的人口将是多少。只需绘制一条延长线，它与之前的线有着相同的趋势。这条延长线与表示2025年的线的交叉点就是估计值。

2. 估计2025年的人口将是多少？

26

14

Excel 提高

正如特雷莎发现的那样，在计算机上使用EXCEL做数学运算很方便。除了制作图表，特雷莎可以使用它计算图表中的数据。

特雷莎想知道这些年来住在小镇上人口的年平均数。在EXCEL中计算平均数很容易，甚至不必用计算器。

下面是特雷莎的EXCEL表，标记了行和列：

	A	B	C
1	年份	人口	平均人口
2	1750	45	
3	1775	231	
4	1800	879	
5	1825	1670	
6	1850	4006	
7	1875	7096	
8	1900	10213	
9	1925	10607	
10	1950	10890	
11	1975	9992	
12	2000	8863	
13	2025	?	

要在EXCEL中进行计算，就必须知道如何编写函数。函数用来计算数据并得到一个答案。在数学中，平均数函数是将所有求平均的数相加然后除以这些数的个数。EXCEL知道如何计算平均数，所以只需告诉EXCEL你想要平均数。

特雷莎在C列的一个空单元格里输入=AVERAGE（B2：B12）。AVERAGE一词告诉EXCEL求平均数，还需要说明哪些数字需要求平均——这些数字从单元格B2到B12。

1. 1750年至2000年期间的平均人口是多少？可以使用EXCEL或计算器(或使用你的大脑)。

2. 如果只想要1900年至2000年的平均人口，你会输入什么？

特雷莎发现了不同年份人口之间的差别。她想看看2000年比1750年人口增加了多少，于是在一个空单元格中输入：

$$=B12-B2$$

3. 如果想知道1950年和1850年的人口差，应该输入什么？

15

小 结

就学计算机而言，搬到一个新的城镇是特雷莎新的开始。她了解了很多计算机知识，如互联网速度、二进制数、编程，等等。你是否能记得一些她学过的知识。

1. 互联网速度是30/4 Mb/s。

下载一个11MB的文件需要多久？那么上传呢？

2. 1 TB 等于多少比特？

3. 邮箱存储空间为20.5MB，已经使用了35%，还剩多少MB的存储空间？

4. 二进制数1011010011是多少？

5. 在"与"命令操作中，如果A=1，B=0，那么Q等于多少？

在"或"命令操作中，对于同样的输入，Q等于多少？

6. 890 × 720的网页像素是多少？

某人的屏幕是640 × 480，他能方便地浏览这个网页吗？为什么？

7. 十六进制数F2193A的十进制数是多少？

这个颜色最接近于红色、绿/黄色，还是蓝色？为什么？

8. 原始文本框为2.4英寸宽、7.5英寸高，按比例重新调整为3.4英寸宽。

调整后的文本框有多高？

参考答案

1.

1. 512/8=64KB/s，980KB/64KB/s=15.31s
2. 都超过1秒，你或许能注意到差别

2.

1. 175.54 s（560 ÷ 8 = 70KB/s， 12 × 1024=12288KB， 12288 ÷ 70=175.54s）
   48 s（2 ÷ 8 = 0.25 MB/s， 12 ÷ 0.25 = 48 s）
   96 s（1 ÷ 8 = 0.125 MB/s， 12 ÷ 0.125 = 96 s）
   768 s（128 ÷ 8 = 16 KB/s， 12 × 1024 = 12288 KB， 12288 ÷ 16 = 768 s）
   19.2 s（5 ÷ 8 = 0.625 MB/s， 12 ÷ 0.625 = 19.2 s）
2. 标准包：10.67/160 s
   特级包：5.33/32 s
   超级包：3.2/32 s
3. 他们应该选特级包；对于下载速度他们应该不会觉察太明显的差别

3.

1. 1 GB × 1024 MB × 1024 KB = 1048576 KB
2. 3 × 1024 × 1024=3145728TB
3. 89 × 8 = 712 比特

4.

1. 0.80 × 10.1 = 8.08 GB（使用的空间）；10.1 − 8.08 = 2.02 GB（剩下的空间）
2. 100 × x = 80 × 10.1， x = 8.08 GB（使用的空间）；是的，一样
3. 8.08 × 1024 = 8273.92 KB
4. 1924 ÷ 1024 = 1.88 GB， 8.08 − 1.88 = 6.2 GB；6.2/10.1 = x/100， x = 61.39%

5.

1. $2 \times 2 \times 2 \times 2 \times 2 \times 2 \times 2 = 128$
2. $2^4 + 0 + 0 + 0 = 8$, $2^4 + 0 + 0 + 2^0 = 9$
3. $2^7 + 2^6 + 0 + 0 + 2^3 + 0 + 0 + 0 = 128 + 64 + 8 = 200$

6.

1. 65, 66, 67
2. 84, 101, 114, 101, 115, 97
3. H = 01001000

7.

1. 9
2. 26
3. 11000001

8.

1. 若 A=1 与 B=1, 则 Q=1
2. A B Q
   0 0 0 若 A=0 与 B=0, 则 Q=0
   0 1 1 若 A=0 与 B=1, 则 Q=1
   1 0 1 若 A=1 与 B=0, 则 Q=1
   1 1 1 若 A=1 与 B=1, 则 Q=1
3. A=1, B=0

9.

```
Program
Number (xCubed)
Print "This program will find the result of the equation
2^3."
   (xCubed= 2 * 2 * 2 )
Print "Your number cubed is ",  (xCubed)
End Program
```

33

10.

1. 307200像素
2. 1280 × 800; 大237568
3. 1024 × 768，因为她的大部分朋友都使用这个分辨率

11.

1. $1 \times 16^1 + 4 \times 16^0 = 20$
2. 红色
3. 接近白色，因为数值很高
4. $12 \times 165 + 0 \times 164 + 3 \times 163 + 0 \times 162 + 15 \times 161 + 15 \times 160 = 12595456$

12.

1. 10英寸
2. $3/6 = 2/X$，$X = 4$英寸
3. A，D

13.

1. x轴代表年份，y轴代表人口
2. 大约8000

14.

1. 5862.91
2. =AVERAGE(B8:B12)
3. =B10-B6

15.

1. $30/8 = 3.75$ MB/s; 11MB/3.75MB/s = 2.93s
   4/8=0.5MB/s; 11MB/0.5MB/s=22s
2. $1 \times 1024 \times 1024 \times 1024 \times 1024 \times 8 = 8796093022208$
3. $0.35 \times 20.5 = 7.175$，$20.5 - 7.175 = 13.325$MB

4. 723
5. 0; 1
6. 640800; 不能，因为屏幕像素低于那个网页像素
7. 15866170; 红色，因为代表红色的那组数值最高
8. $2.4/7.5 = 3.4/X$, $X = 10.625$ 英寸

INTRODUCTION

How would you define math? It's not as easy as you might think. We know math has to do with numbers. We often think of it as a part, if not the basis, for the sciences, especially natural science, engineering, and medicine. When we think of math, most of us imagine equations and blackboards, formulas and textbooks.

But math is actually far bigger than that. Think about examples like Polykleitos, the fifth-century Greek sculptor, who used math to sculpt the "perfect" male nude. Or remember Leonardo da Vinci? He used geometry—what he called "golden rectangles," rectangles whose dimensions were visually pleasing—to create his famous *Mona Lisa*.

Math and art? Yes, exactly! Mathematics is essential to disciplines as diverse as medicine and the fine arts. Counting, calculation, measurement, and the study of shapes and the motions of physical objects: all these are woven into music and games, science and architecture. In fact, math developed out of everyday necessity, as a way to talk about the world around us. Math gives us a way to perceive the real world—and then allows us to manipulate the world in practical ways.

For example, as soon as two people come together to build something, they need a language to talk about the materials they'll be working with and the object that they would like to build. Imagine trying to build something—anything—without a ruler, without any way of telling someone else a measurement, or even without being able to communicate what the thing will look like when it's done!

The truth is: We use math every day, even when we don't realize that we are. We use it when we go shopping, when we play sports, when we look at the clock, when we travel, when we run a business, and even when we cook. Whether we realize it or not, we use it in countless other ordinary activities as well. Math is pretty much a 24/7 activity!

And yet lots of us think we hate math. We imagine math as the practice of dusty, old college professors writing out calculations endlessly. We have this idea in our heads that math has nothing to do with real life, and we tell ourselves that it's something we don't need to worry about outside of math class, out there in the real world.

But here's the reality: Math helps us do better in many areas of life. Adults who don't understand basic math applications run into lots of problems. The Federal Reserve, for example, found that people who went bankrupt had an average of one and a half times more debt than their income—in other words, if they were making $24,000 per year, they had an average debt of $36,000. There's a basic subtraction problem there that should have told them they were in trouble long before they had to file for bankruptcy!

As an adult, your career—whatever it is—will depend in part on your ability to calculate mathematically. Without math skills, you won't be able to become a scientist or a nurse, an engineer or a computer specialist. You won't be able to get a business degree—or work as a waitress, a construction worker, or at a checkout counter.

Every kind of sport requires math too. From scoring to strategy, you need to understand math—so whether you want to watch a football game on television or become a first-class athlete yourself, math skills will improve your experience.

And then there's the world of computers. All businesses today—from farmers to factories, from restaurants to hair salons—have at least one computer. Gigabytes, data, spreadsheets, and programming all require math comprehension. Sure, there are a lot of automated math functions you can use on your computer, but you need to be able to understand how to use them, and you need to be able to understand the results.

This kind of math is a skill we realize we need only when we are in a situation where we are required to do a quick calculation. Then we sometimes end up scratching our heads, not quite sure how to apply the math we learned in school to the real-life scenario. The books in this series will give you practice applying math to real-life situations, so that you can be ahead of the game. They'll get you started—but to learn more, you'll have to pay attention in math class and do your homework. There's no way around that.

But for the rest of your life—pretty much 24/7—you'll be glad you did!

1
DOWNLOAD SPEEDS

Teresa just moved to a new city and a new house. She has picked out her new room, and her parents are decorating the rest of the house.

When they first moved in, they didn't have any computers set up. Then when Teresa's brother AJ unpacked his laptop, and her dad set up the family's computer, they discovered they didn't have Internet access. With all the things they had to do for the move, Teresa's parents had forgotten to get Internet service for their new home.

Teresa and her dad go to the local library where there are plenty of computers and free Internet. They look up what the best Internet deals are, so they can call and order an Internet package.

Teresa sees a lot of numbers she doesn't immediately understand. What does 512/128 Kb/s mean, for example? Her dad explains that those are the download and upload speeds. The first number shows how fast you can transfer information from another computer system onto your own. You can download music, spreadsheets, e-mail attachments, and a lot more.

You'll first need to understand how people write download speeds. Unfortunately, there isn't really a standard notation. People commonly use Kb/s for kilobits per second or Kbit/s. Megabits per second could be Mb/s or Mbit/s. They all mean the same thing.

If you are trying to download a file that is 67 megabits, you would be able to download it much faster with a higher speed, like 6 Mb/s. The download will be slower if your Internet download speed is 512 Kb/s.

Here are the steps you need to take to figure out just how long a file will take to download.

- Convert your download speed from kilobits or megabits per second to kilobytes or megabytes per second by dividing by 8. (This step will be explained more in section 3.)
- Make sure the file size is in the same units as the download speed. If it's not, use the information below to convert the numbers into the same unit of speed.

1024 bytes = 1 kilobyte
1024 kilobytes = 1 megabyte
1024 megabytes = 1 gigabyte

- Divide the file size by the download speed to get the number of seconds the download will take.

Now find how long a 980-kilobyte file would take to download at 6 Mb/s:
6 Mb/s ÷ 8 = .75 megabytes per second
980 kilobytes ÷ 1024 = .96 megabytes
.96 megabytes ÷ .75 megabytes per second = 1.28 seconds

1. How long would it take the file to download at 512 Kb/s?

2. Are either download speeds more than a second? Do you think you would notice the difference between the two speeds?

2
UPLOAD
SPEEDS

Now that Teresa understands download speeds, she thinks she also gets upload speeds. Remember the second number in the Internet package offers? The second number in 512/128 Kb/s is the upload speed. Uploading means to transfer from your computer to another computer system. You can upload pictures to Facebook or videos onto YouTube.

Teresa and her dad need to figure out what upload speed they want. Once they decide, they can choose the right Internet package for them.

Upload speeds tend to be slower than download speeds. But, you can find how long it would take you to upload files at different upload speeds the same way you do with download speeds.

1. If you are trying to upload a file that is 12 megabytes, how long would it take in seconds to upload with the following upload speeds:

    560 Kb/s:

    2 Mb/s:

    1 Mb/s:

    128 Kb/s:

    5 Mb/s:

Now that Teresa can see that faster upload and download speeds mean she can add music to the computer faster and watch TV faster, she agrees with her dad that they shouldn't get the slowest Internet package. But how fast should they get?

Here are their choices:

standard: 15/1 Mb/s, $40/month
extra: 30/5 Mb/s, $50/month
ultra: 50/5 Mb/s, $65/month

The very biggest files anyone will probably be downloading or uploading on the computer will be about 20 megabytes, which is fairly large, because Teresa's mom sometimes does some graphic design work.

2. How fast will 20 megabytes upload and download in each Internet package?

standard:
extra:
ultra:

3. Teresa's parents don't really want to spend more than $50 a month. Which Internet package do you think they should get? Do you think they will notice the loss in speed if they don't order the ultra package?

3
COMPUTER STORAGE

Teresa and her dad go home, but it's too late to call the Internet company tonight—they're closed. Teresa sits down at the computer, but she doesn't really know what to do. She's used to surfing the Internet or watching her favorite TV shows online.

She clicks around, exploring all the buttons she's never used on the computer. At one point, she comes across a screen that shows her how much storage the computer has. She sees the words megabytes and gigabytes. They look awfully similar to the kilobits and megabits she learned about with download and upload speeds. She calls her brother AJ over, who explains that bytes, kilobytes, and megabytes are used to describe how much storage space a computer has. Unlike Internet speeds, which are given in bits, computer storage is expressed in bytes. Luckily, the two are related.

A bit is the smallest unit of information on a computer. A byte is just a little bigger—1 byte is actually 8 bits.

You can talk about bigger pieces of information on a computer, too. Kilobytes are bigger than bytes, megabytes are bigger than kilobytes, and so on. The chart below shows how all these units of measurement relate to each other:

8 bits = 1 byte
1024 bytes = 1 kilobyte
1024 kilobytes = 1 megabyte
1024 megabytes = 1 gigabyte
1024 gigabytes = 1 terabyte
1024 terabytes = 1 petabyte
1024 petabytes = 1 exabyte

1. How many kilobytes are in 1 gigabyte?

2. How many terabytes are in 3 exabytes?

The numbers can get pretty big!

When you are talking about a kilobit or megabit in Internet speed, you're really talking about a kilobyte or a megabyte divided by 8.

3. How many bits are in 89 bytes?

4
E-MAIL
STORAGE

When Teresa's family sets up their Internet, the first thing Teresa does is check her e-mail. She's wondering if any of her friends from her old town have e-mailed her.

After she reads all her new e-mails, she notices a little bar in the corner of the screen that says 80% full. She has never really noticed it before, so she looks closer.

Sure enough, she also sees that it says "80% of your 10.1 GB full." More gigabytes! She already knows what that means in terms of how much space she has on her computer. Now she realizes her e-mail also has storage space for all her e-mails and e-mail attachments. On the next page, figure out how much storage space Teresa has used and how much she has left.

Teresa's e-mail program gives her 10.1 gigabytes of storage. To figure out how many gigabytes she has left, you'll have to work with percentages.

To find out how much 80% of 10.1 gigabytes is, convert the percentage into a decimal number by moving the decimal point two spaces to the left. Then 80% becomes .80. Multiply the decimal number by the total number of gigabytes, which tells you how much space Teresa has already used.

1. How much storage space has she used in gigabytes? How much space does she have left?

You can also turn the percentage into a fraction, to find out the answer another way. Put the percentage value over 100.

$$80\% = \frac{80}{100}$$

Then set the fraction equal to the number of gigabytes used in Teresa's e-mail, divided by the total number of gigabytes she started with. Use an X for the number of gigabytes she has used, because you don't know that number yet.

$$\frac{80}{100} = \frac{x}{10.1}$$

2. What do you get by cross-multiplying? Is it the same answer as before?

3. How many kilobytes has she used?

Teresa thinks she should delete some of her old e-mails to make more room for new e-mails. She goes through her really old e-mails and deletes 1924 megabytes worth of stuff.

4. How many gigabytes has her e-mail used up now? What percentage of space does she have now?

5
BINARY CODE

Teresa is wondering what bits and bytes and gigabytes actually are. She understands that they're ways to measure information in a computer, but what exactly are they measuring? How big is a bit to begin with? She asks AJ all these questions.

AJ tells her that she needs to understand binary code, the language that computers use. Computers can't easily understand the letters and numbers that humans use, so instead they store information as strings of 1s and 0s. Binary means "having only two states," and the binary code that computers use only has 1s or 0s.

The computer uses different combinations of these 1s and 0s to represent other characters. We'll talk more about that in chapter 6.

The word bit actually is short for binary digit. And one binary digit is the very smallest piece of computer language. A binary digit in computer language is a 0 or 1. You could say that 1 equals "on" or "yes," and 0 equals "off" or "no."

Counting looks different in binary than in our usual system for counting. We normally use a 10-digit system, made up of 10 digits (0 through 9). You can only count with two digits in binary code—but you need to count a lot higher than 2!

First, you need to be familiar with the idea of the powers of two. A power is a way of saying how many times you multiply a number by itself. Two to the power of 0 looks like 2^0 and equals 1. Two to the power of 1 looks like 2^1 and equals 2. Here are some more:

$$2^2 = 2 \times 2 = 4$$
$$2^3 = 2 \times 2 \times 2 = 8$$
$$2^4 = 2 \times 2 \times 2 \times 2 = 16$$

1. What is 2^7?

Here are the numbers 0 through 10 as they would be written in binary:

0000 (= 0)   0110 (= 6)
0001 (= 1)   0111 (= 7)
0010 (= 2)   1000 (= 8)
0011 (= 3)   1001 (= 9)
0100 (= 4)   1010 (= 10)
0101 (= 5)

To count in the binary system, you always start with 0. Next is a 1. But what do you do after that? Think back to powers of 2. Every digit you see can be assigned a power of 2, and these can be switched "on," when they have a 1, or "off," when they have a 0. Then you add together all the columns that have 1s in them. So for 10 (= 2 in our normal way of counting), you divide 10 into two columns. The column on the right is 2^0 and the column on the left is 2^1. Only the left column is turned "on," because it is a 1. That is the only column you count. You know that 2^1 is 2, and that is your answer.

For 3, both columns are now turned on. From right to left, you have $2^0 + 2^1 = 3$. For 4, you now have a 2^3 column. Only that column is turned on, and $2^3 = 4$.

2. How would you explain the number 8? And 9?

There are 8 bits in a byte. With 8 bits, values in a byte can range from 00000000 to 11111111. (There are 8 options for a 1 or 0.)

3. What value would the byte 11001000 be?

6
WRITING IN
BINARY

AJ continues to explain more about binary, bits, and bytes, and what it all actually means. He asks Teresa to open up Word Pad, the most basic text program on their computer. He asks her to type out her name, which she does.

"What do you see?" he asks her. Teresa says she just sees the letters that make up her name. But AJ says that to the computer, the letters look like binary code. The computer can only read 1s and 0s, not letters. The computer reads each character on the screen as 1 byte (8 bits). Then he shows her how to write in binary, without using any letters! Learn how on the next pages.

Each character (letter or symbol, like a period or dash) on the screen is 1 byte of information to a computer.

People have come up with a system where each character is assigned a particular byte code. For example:

A = 01000001
B = 01000010
C = 01000011

1. What is A, B, and C if you convert it into normal (decimal) numbers?

You need to know which numbers—in either binary or decimal—match up with which characters to be able to write in binary.

When Teresa writes out her name, the bytes of information in binary code she is using are:

T = 01010100
E = 01100101
R = 01110010
E = 01100101

S = 01110011
A = 01100001

2. What is each letter in decimal numbers?

3. If Teresa's last name starts with the byte represented by 72, which of the following letters is that?

M = 01001101
D = 01000100
H = 01001000

7
CONVERTING
TO
HEXADECIMAL

AJ tells Teresa that there's another kind of number that computers understand other than binary: hexadecimal. Just like binary numbers are base 2 and the normal numbers we use are base 10, hexadecimal numbers are base 16. This means that there are more digits than we have numerals for, since our writing system only has ten numerals! The computer starts using letters after 9, so it has enough symbols. Here's how you write the numbers 1 to 16 in hexadecimal:

1 = 1	9 = 9
2 = 2	A = 10
3 = 3	B = 11
4 = 4	C = 12
5 = 5	D = 13
6 = 6	E = 14
7 = 7	F = 15
8 = 8	10 = 16

You can convert between binary and hexadecimal pretty easily. Since 16 is the same as 2^4, every four digits of binary works out to one digit of hexadecimal.

Binary	Decimal	Hexadecimal
0001	1	1
0101	5	5
1010	10	A
1011	11	B
1100	12	C
1101	13	D
1111	15	F

1. What's the binary number 1001 in hexadecimal?

2. What's the hexadecimal number 1A in base 10?

3. What's the hexadecimal number C1 in binary?

8
COMPUTER LOGIC

Soon after she starts at her new school, Teresa has to do a research. She has to research her town's history. She doesn't really know much about her new town, so she has to do a lot of research.

She starts by using Google. First, she searches for "Brookeville," the name of her town. She gets a lot of results, but most of them don't have anything to do with her town. She wonders how to search so she only gets the results she wants. Her project would be a lot easier if she could narrow down her search results.

Luckily, there is an easy way to do that. She can use a Boolean search. Boolean logic involves the terms "and," "or," and "not." Boolean searches are part of a system of logic that computers use all the time. You can use it for Google searches, but computers use it to do everyday functions too.

First, Teresa could just add another term to her search. Brookeville is in New York State, so she could search for "Brookeville New York." She doesn't have to put the "and" in there, because Google assumes you mean "and." This search will produce results that contain both the words Brookeville and New York.

Teresa notices a lot of her results are for Brookevilles in other states, especially Texas. She could search for "Brookeville –Texas." The minus sign before Texas means "not" in Google search terms. Now, she leaves out results with the word Texas.

The town also used to be called Brooketon a long time ago. Teresa can search for both Brookeville and Brooketon at once by searching for "Brookeville OR Brooketon." The "or" returns results that contain either or the other.

Computers use this logic too, just not in word form. They use 1s and 0s, of course. You can use tables to show the logic. Here is a table for an AND command, and the explanation. In this table, A is the first input, B is the second input, and Q is the output. The only way for Q to be 1 is for both A and B to be 1.

A B Q
0 0 0 *If A is 0 AND B is 0, Q is 0.*
0 1 0 *If A is 0 AND B is 1, Q is 0.*
1 0 0 *If A is 1 AND B is 0, Q is 0.*
1 1 1

1. What would 1 1 1 mean in an AND table?

   The OR table looks like this. If either A or B is 1, the output will be 1. Fill in the rest of the explanations.

2. **A B Q**
   0 0 0 If A is 0 and B is 0, Q is 0.
   0 1 1
   1 0 1
   1 1 1

   Here is the NOT table. There is only one input, and the output is always the opposite.

   A Q
   0 1
   1 0

3. What would Teresa's "Brooketon –Texas" search look like in computer terms? You can think of it as searching for "Brooketon AND NOT Texas," which eliminates results from a Brooketon AND Texas search.

9
BASIC PROGRAMMING

Teresa is starting to make some good friends at her new school. One of her new friends, Mel, happens to be really good at computers.

One day, when Mel is over at Teresa's house, Teresa mentions her calculator ran out of batteries and she hasn't had time to get new ones yet. Mel tells her she shouldn't worry about her calculator because she can just turn her computer into a calculator for math.

Teresa will need to learn a little programming first, Mel explains. Programming is how you tell computers commands, kind of like another language you need to know to work with computers. The computer will translate the programming commands, called programming code, into binary numbers, so the computer recognizes what you're telling it to do. Mel has already studied programming a little bit, so she shows Teresa what to do.

There are hundreds of different programming languages. It doesn't really matter which one you use, as long as you follow the rules for that language. All of the examples in this section are a made-up language that could exist. The computer languages you know or learn will look a little different, but they'll do the same things.

Mel tells Teresa that all computer programs are divided into coding lines. Each line has one command. Programs have to start by telling the computer a program is running and that it is ending, like this:

```
1 Program
2 End Program
```

What you put between the two commands are the instructions for the computer to become a calculator, or run whatever program you want.

Teresa's math homework involves a lot of cubed numbers, which are numbers times themselves three times. Think back to binary numbers and powers of 2. Two cubed is 23, or 2 x 2 x 2. Teresa wants her computer program to calculate numbers cubed.

Here are the steps. Remember, the first is "Program" and the last is "End Program":

First, they have to tell the computer to make space to store the number they are cubing and the number cubed, before they even writes the code for cubing. This is called declaring a variable. Our variable will be called "xCubed." You can make a new variable by deciding what kind of information will be stored in it (text or a number) and naming the variable. For example, "Number (xCubed)" will make a new number variable named xCubed.

Next, Mel tells Teresa she will have to tell the computer to cube the number 2 by multiplying it three times, and put the resulting value into "xCubed."

Finally, Teresa has to tell the computer to give the answer to the cube equation in the seventh line. She can have it print out the answer in words.

Rearrange these lines of computer code in the order you think they should go, to tell the computer how to calculate square roots. The words not in parentheses are the commands, the symbols in parentheses modify the commands.

Print "This program will find the result of the equation 23."
End Program
Program
Number (xCubed)
Print "Your number cubed is ", (xCubed)
(xCubed= 2 * 2 * 2)

10
WEBSITE DESIGN: PIXELS

Now Teresa wants to create her own website to communicate with her friends in her old town. She's so busy all the time that she doesn't always have time to call or e-mail them all individually. A website with updates about herself would be a good way to stay in touch with all her friends at once.

Teresa is using a website-building program that makes it a little easier. She still has to make decisions about what she wants her site to look like. For example, she needs to pick the right dimensions. She doesn't want her site to be too big for the page, or too small.

Websites are measured in pixels, just like other things are measured in inches, centimeters, or miles. One pixel is a tiny dot that makes up an image on a computer screen. One image is made up of thousands or millions of pixels. The pixels are so small that you can't see them individually—they all blend together into a smooth picture. A good website will have the right dimensions, so it looks nice.

When you pick the number of pixels you want a website to be, you choose between options that look like this:

$$640 \times 480$$
$$1024 \times 768$$
$$800 \times 600$$

In other words, a website that is 640 x 480 is 640 pixels across the width of the screen, and 480 pixels along the height. This number is called the resolution.

You calculate the number of pixels on that website the same way you calculate area, by multiplying the length and the width.

1. How many pixels are in a 640 x 480 website?

2. Which has more pixels, a website that is 1024 x 768 or 1280 x 800? How many more?

A good website will have resolutions that match the most people's computer screens. If your computer's resolution is 800 x 600, but you're looking at a website that is 1024 x 768, the website is bigger. You won't be able to see it all on one screen. You'll have to scroll to the right to see the whole website.

Teresa takes a poll of her friends. She asks them what resolution their computer screens are, so she can pick the right resolution for her website. These are the results:

$$800 \times 600 = 2 \text{ people}$$
$$1024 \times 768 = 6 \text{ people}$$
$$1280 \times 800 = 1 \text{ person}$$

3. Which computer screen resolution should Teresa try to match? Why?

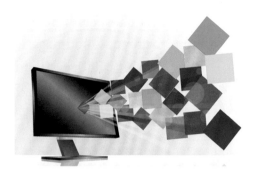

11
WEBSITE DE-SIGN: COLORS

Next, Teresa has to pick some colors for her website. As she's picking colors on the website design program, she notices that each color has a funny number that identifies it. The number actually contains a bunch of numerals, but also some letters. She notices a purplish color, for example, is C030FF. What does it mean?

Teresa is looking at a way of counting called hexadecimal. Like binary, hexadecimal is just a different way of thinking about counting. We normally use a system called decimal, based on the number 10. Binary is based on the number 2. Hexadecimal is based on 16. It has the ten numerals 0 through 9. And instead of making up new symbols for the rest of the numbers, hexadecimal uses the six letters A through F.

To get the hexadecimal code for a color, you're really combining 3 different hexadecimal numbers. The C0 in the purple is how much red is in the color. The 30 is how much green there is. And FF is how much blue there is.

You will need to better understand how to count in hexadecimal first.

The first 16 numbers in hexadecimal are 0, 1, 2, 3, 4, 5, 6, 7, 8, 9, A, B, C, D, E, F.

After that, is:

$$10 \ (= 16)$$
$$11 \ (= 17)$$
$$12 \ (= 18)$$

What is going on? Remember back to binary, when you learned that each column in the number equals 2 to a power. The same is true of hexadecimal, except that you are working with powers of 16, not powers of 2.

10 really means $1 \times 16^1 + 0 \times 16^0$, which equals 16.

1. What does 14 equal, using powers of 16?

You will need to remember which values A through F equal.

The number 4C, is really $4 \times 16^1 + 12 \times 16^0$. C is 12 in hexadecimal. So 4C equals 64 + 12, or 76.

For colors, computers always mix together red, green, and blue, in that order. Green already contains yellow, which is how the computer can get yellow. Each color has a two-digit hexadecimal code. That's why each color actually has 6 digits, because it is combining 3 colors. The first two numbers tell you how much red is in the color, the second set of 2 numbers tell you how much green/yellow there is, and the third set of 2 numbers tells you how much blue there is.

2. What color would you expect E41D14 to be?

The lower the number, the darker (closer to black it is). The higher the number, the closer to white it is: 000000 is black, while FFFFFF is white.

3. Would the blue value E9 be closer to black or white? Why?

That purple color Teresa chose was C030FF. Now you have to deal with powers of 16, 0 through 5, because there are 5 columns of numbers.

4. What is the purple color's value in the decimal system?

12
WEBSITE DESIGN: PRO-PORTION

Teresa's website is halfway done, but she still has to add images and text boxes to it. She has a whole list of things she wants to include, but she has to make sure it all fits on each webpage. To fit everything, she has to play around with the size of each image or text box. Teresa quickly discovers that she is dealing with proportions. When she changes an image size, for example, she could choose to change it proportionally. As she makes the image wider, she also makes it longer, which keeps the image looking normal, just bigger. She could also change it disproportionally, by making it too wide or too long. That makes it look squished. Understanding proportions makes website design easier, and makes websites look a lot nicer.

Two things that are proportional have the same relationship. A picture that is 3 inches by 6 inches has a specific relationship between its length (3 inches) and height (6 inches). You can give that relationship as a fraction:

$$\frac{\text{width}}{\text{length}}$$

$$\frac{3 \text{ inches}}{6 \text{ inches}}$$

Teresa wants the pictures she's putting on the website to be larger than 3 inches by 6 inches. However, she wants it to look the same and have the same proportions, rather than get squished.

In fact, she wants the picture to be 5 inches in width. To find the new length, you have to find the fraction that has the same relationship between width and length as the original picture. You already know that relationship is ⅜.

Set the two picture size proportions equal to each other, leaving out the missing information. Add an X for the measurement you don't know.

$$\frac{3}{6} = \frac{5}{X}$$

Now cross-multiply:

1. $3 \times X = 5 \times 6$

   $X =$

2. If Teresa wanted to make the height of the text box an inch smaller, what would the new, proportional width and height be?

3. Which of these screen resolutions is proportional to 1024 x 768? (There may be more than one answer.)

   A. 960 x 720
   B. 976 x 580
   C. 1080 x 925
   D. 1140 x 855

13
EXCEL

Teresa's computer has Excel on it, a program she uses from time to time for school projects. She could even use Excel for that history project she was working on before, about the history of her new town.

The second part of her project asks how many people have lived in the town over the years. Teresa could just do the research and write down all the numbers she finds for each year, but she wants to take it a step further. She decides to type all her information into Excel and make it into a graph representing the population of her town. See what the chart and graph look like on the next page.

This is what Teresa types into an Excel chart, which will be made into a graph.

Year	Population
1750	45
1775	231
1800	879
1825	1,670
1850	4,006
1875	7,096
1900	10,213
1925	10,607
1950	10,890
1975	9,992
2000	8,863
2025	?

1. What should the x-axis on Teresa's graph be labeled? What about the y-axis?

Here is what the graph looks like. Add in the axis labels.

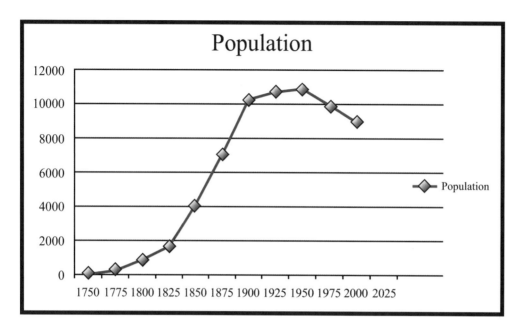

Population

You can **estimate** what the 2025 population will be with this graph. Just draw or imagine a line extending out with the same **slope** as what has come right before. The point at which your line crosses the year is your estimate.

2. What do you estimate the population will be in 2025?

14
MORE EXCEL

Excel is a great program for doing math on a computer, as Teresa is discovering. Besides making graphs, Teresa can use it to make calculations with the **data** she has entered into a chart.

Teresa wants to find the average number of people who have lived in her town over the years. Finding the average is easy on Excel, and Teresa won't even have to pick up a calculator.

Here is what Teresa's Excel chart looks like, with the columns and rows labeled:

	A	B	C
1	Year	Year	Population
2	1750	45	
3	1775	231	
4	1800	879	
5	1825	1,670	
6	1850	4,006	
7	1875	7,096	
8	1900	10,213	
9	1925	10,607	
10	1950	10,890	
11	1975	9,992	
12	2000	8,863	
13	2025	?	

To do calculations in Excel, you have to know how to write functions. Functions are just things you do to data to get an answer. In math, the function for average is adding up all the numbers you're averaging and dividing by how many numbers you added. Excel already knows how to average numbers, so all you have to do is tell Excel you want an average.

Teresa types in =AVERAGE(B2:B12) in an empty cell in column C. The word AVERAGE tells Excel she wants to find the average. She also has to tell it what numbers she wants to average. Those numbers are in cells B2 to B12.

1. What is the average population between 1750 and 2000? You can use Excel or a calculator (or do it in your head).

2. What would you type if you wanted to only find the average of the years 1900 to 2000?

Teresa could also find the difference between different populations. If she wanted to see how much the population had grown from 1750 to 2000, she would type this into an empty cell:

=B12 – B2

3. What would she type if she wanted to know the difference between the years 1950 and 1850?

15
PUTTING IT
ALL TOGETHER

Moving to a new town was a new start for Teresa, as far as computers are concerned. She has learned a lot about Internet speed, binary numbers, programming, and more. See if you can remember some of what she has learned.

1. Your Internet speed is 30mbs/4 mbs.

   How long would it take a file that is 11 megabytes to download?

   How long would it take the file to upload?

2. How many bits are in a terabyte?

3. You have used up 35% of your e-mail program's storage of 20.5 megabytes. How many megabytes of storage do you have left?

4. What is the binary number 1011010011?

5. In an AND computer operation, what will Q be if A is 1 and B is 0?

   What will Q be with the same inputs, but an OR operation?

6. How many pixels are on a website that is 890 x 720?

Will someone with a 640 x 480-sized screen be able to see the website well? Why or why not?

7. What is the hexadecimal number F2193A in decimal form?

   What color is that closest to—red, green/yellow, or blue? Why?

8. You are resizing a text box that was originally 2.4 inches wide and 7.5 inches long. You want it to stay proportional, but be 3.4 inches wide.

   How long will the text box be?

ANSWERS

1.

1. 512/ 8 = 64 kilobytes per second, 980 kilobytes/64 kilobytes per second = 15.31 seconds
2. Both are more than a second; you would likely notice the difference.

2.

1. 175.54 seconds (560 ÷ 8 = 70 kilobytes per second, 12 x 1024 = 12288 kilobytes, 12288 ÷ 70 = 175.54 seconds)
   48 seconds (2 ÷ 8 = .25 megabytes per second, 12 ÷.25 = 48 seconds)
   96 seconds (1 ÷ 8 = .125 megabytes per second, 12 ÷.125 = 96 seconds)
   768 seconds (128 ÷ 8 = 16 kilobytes per second, 12 x 1024 = 12288 kilobytes, 12288 ÷ 16 = 768 seconds)
   19.2 seconds (5 ÷ 8 = .625 megabytes per second, 12 ÷.625 = 19.2 seconds)
2. Standard: 10.67 ÷ 160 seconds
   Extra: 5.33 ÷ 32 seconds
   Ultra: 3.2 ÷ 32 seconds
3. They should choose the extra package; they will probably not notice the difference too much in download speeds.

3.

1. 1 gigabyte x 1024 megabytes x 1024 kilobytes = 1,048,576 kilobytes
2. 3 exabytes x 1024 petabytes x 1024 terabytes = 3,145,728
3. 89 x 8 = 712 bits

4.

1. .80 x 10.1 = 8.08 gigabytes used; 10.1 – 8.08 = 2.02 gigabytes left
2. 100 x X = 80 x 10.1, X = 8.08 gigabytes left; yes, same answer
3. 8.08 x 1024 = 8273.92 kilobytes
4. 1924 ÷ 1024 = 1.88 gigabytes, 8.08 – 1.88 = 6.2 gigabytes; 6.2/10.1 = X/100, X = 61.39%

5.

1. 2 x 2 x 2 x 2 x 2 x 2 x 2 = 128
2. $2^4 + 0 + 0 + 0 = 8$, $2^4 + 0 + 0 + 2^0 = 9$
3. $2^7 + 2^6 + 0 + 0 + 2^3 + 0 + 0 + 0 = 128 + 64 + 8 = 200$

6.

1. 65, 66, 67
2. 84, 101, 114, 101, 115, 97
3. H = 01001000

7.

1. 9
2. 26
3. 11000001

8.

1. If A is 1 AND B is 1, Q is 1.
2. A B Q
   0 0 0 If A is 0 and B is 0, Q is 0.
   0 1 1 If A is 0 and B is 1, Q is 1.

1 0 1 If A is 1 and B is 0, Q is 1
1 1 1 If A is 1 and B is 1, Q is 1
3. A = 1, B = 0

9.

Program
Number (xCubed)
Print "This program will find the result of the equation 2^3."
(xCubed= 2 * 2 * 2)
Print "Your number cubed is ", (xCubed)
End Program

10.

1. 307,200 pixels
2. 1280 x 800; 237,568 more
3. 1024 x 768, because most of her friends have that resolution.

11.

1. $1 \times 16^1 + 4 \times 16^0 = 20$
2. Red
3. Closer to white, because it is a high number.
4. 12 x 165 + 0 x 164 + 3 x 163 + 0 x 162 + 15 x 161 + 15 x 160= 12,595,456

12.

1. 10 inches
2. ⅜ = ⅖X, X =4 inches
3. a, d

13.

1. The x-axis is Year and the y-axis is Population
2. Around 8,000.

14.

1. 5,862.91
2. =AVERAGE(B8:B12)
3. =B10-B6

15.

1. $30/8$ = 3.75 megabytes per second; 11 megabytes/3.75 megabytes per second = 2.93 seconds
   $4/8$ = .5 megabytes per second; 11 megabytes/$.5$ megabytes per second = 22 seconds
2. 1 x 1024 x 1024 x 1024 x 1024 x 8 = 8,796,093,022,208
3. .35 x 20.5 = 7.175, 20.5 – 7.175 = 13.325 megabytes
4. 723
5. 0; 1
6. 640,800; No, because their screen is smaller than the website.
7. 15866170; Red, because the highest values are in the two red columns of the number.
8. $2.4/7.5 = 3.4/X$, X = 10.625 inches